MW00901801

# MATHAPALOOZA

A Collection of Math Poetry for Primary and Intermediate Students

Franny Vergo

*AuthorHouse™*
*1663 Liberty Drive*
*Bloomington, IN 47403*
*www.authorhouse.com*
*Phone: 1-800-839-8640*



*Published by AuthorHouse 5/6/2013*

*ISBN: 978-1-4685-4269-1 (sc)*
*ISBN: 978-1-4685-4268-4 (e)*

*Library of Congress Control Number: 2012900695*

*Mathapalooza* is dedicated to all of my math students, whom I had the privilege of teaching and learning from the past thirty-six years!

# Contents

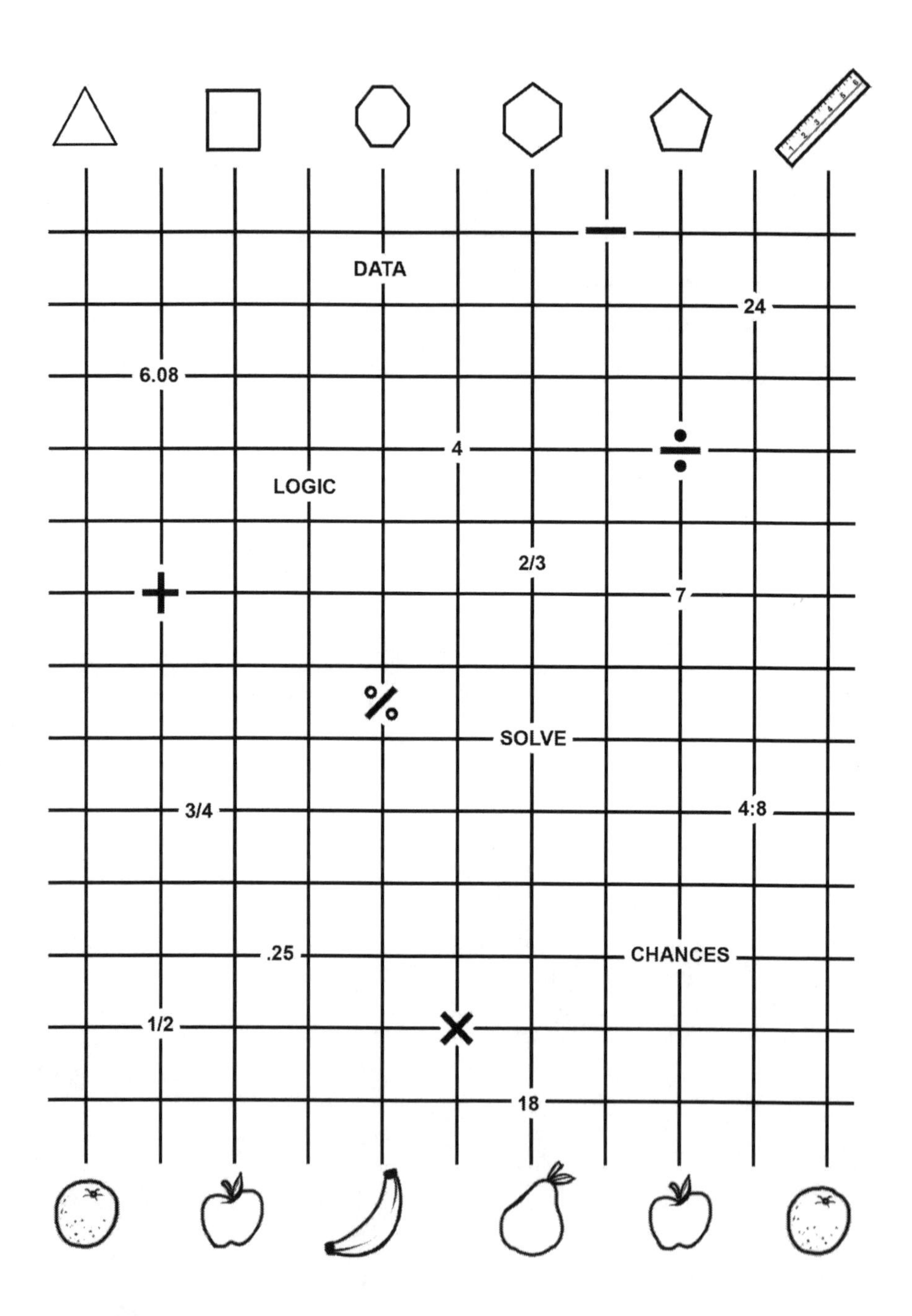
DATA
24
6.08
4
LOGIC
2/3
7
%
SOLVE
3/4
4:8
.25
CHANCES
1/2
18

# MATHEMATIC BRAIN WIZARD

Mathematics is really cool, and it helps enhance your brain,
NUMBERS floating through your head and knowledge that you gain.
ADD, SUBTRACT, MULTIPLY, or sometimes you DIVIDE.
PROBLEM SOLVE and LOGIC PUZZLES: now your brain is on a ride.

DECIMALS, FRACTIONS, and PERCENTS help you with parts of wholes.
Finding some different MEASUREMENTS is one of your math goals.
PROBABILITY is always fun, using spinners, cards, and dice.
In GEOMETRY, draw lots of lines and shapes, and be concise.

Find PATTERNS in a group of numbers or in a picture set.
COLLECT a lot of DATA , and GRAPH the data that you get.
Use RATIOS to compare two different amounts that you see.
Study these math topics, and a mathematic brain wizard you will be.

# TRICKY MATH TRICKS

There are many tricky math tricks that we use a lot.
They help you get the answers that you can quickly jot.

One is using your ten fingers for the times table of nine;
This trick will give you the right nines product every single time.
Just hold up your two hands, with both your palms facing out and fanned,
Bend down the finger of the second factor, starting on your left hand.
Fingers to the left are tens, and fingers to the right are ones.
Now count the fingers up and you've got your answer, and you're done.

Another helps us remember the symbols "less than" or "greater."
The bigger number always gets eaten by the alligator.

Next, a number that is tricky is the digit three. You see,
If a two-digit number's sum breaks evenly by three,
Then the whole number divides by three, and this will always be.

And how about when you add/subtract with even and/or odd?
You always know the answer type, and amazed, you sit and nod,
Because even/even and odd/odd will always give you even,
And odd/even or even/odd equals odd—now you're achievin'.

And here are a couple phrases, whose words remind you of each step.
These cool phrases help a lot, so learn them and you'll be all set.
"Does McDonald's Sell Burgers?" are the four steps for long division:
Divide, Multiply, Subtract, and Bring down the next number, with precision.
"Please Excuse My Dear Aunt Sally" works for the order of operations:
Parentheses, Exponents, Multiply, Divide, Add, and Subtract formations.

# YOU SAY YOU DON'T LIKE MATH?

You say you don't like math?
Well, then, just let me ask you:

Do you like to
Play a lot of games and sports
On the fields or on the courts?
Toss a big six on a die,
Move your piece or move your guy,
Count your points or read your score,
Wear a jersey from the store,
Spin a high ten on a spinner,
Throw your cards down and yell, "Winner!"?

Do you like to
Count the money earned for each chore
And then spend it at a local store?
Listen to your favorite tune,
Learn to play the big bassoon,
Find chapter twenty-one in a book,
Measure ingredients when you cook,
Read three on the clock and time to leave school,
Or four o'clock, time to jump into your pool?

Well, guess what? You were right!
You don't like math. You love it!

# THE VALUE OF EACH PLACE

PLACE VALUE is the basis of all math,
Because it will always guide your path.
It gives you the value of each place
With every number's PERIOD base.

ONES and THOUSANDS, then come MILLIONS and BILLIONS,
The first four periods followed by TRILLIONS.
ONES, TENS, HUNDREDS within each period, you will see,
Together they give the digit's value, and that's the key.

With place value you also need to know
Any PLACE CAN BE HELD with a ZERO.
Then the number gets larger and it will need
Some COMMAS to make it easier to read.

# WHAT IS THE POINT OF THE DECIMAL POINT?

What is the point of the decimal point?. To you I do implore.
We always see it between two numbers, and now we will explore.
If you look closely you will see, it separates whole numbers from parts,
TENTHS, HUNDREDTHS, and THOUSANDTHS, right of the point it all starts.

Tenths and hundredths are the places we use when we write a money amount,
Between the dollar bills and coins, two decimal places we count.
Money goes to hundredths, 'cause one dollar is one hundred cents each time.
Put a zero in the tenths if the coins equal less than a dime.

100
20
TWENTY
20
FIFTY
50
50
FIFTY
1
DOLLAR
10
TEN
1
TEN
10
FIVE
5
10
CENTS
5
CENTS
5
FIVE
ONE
2
1
TWO
1
25
CENTS
ONE
50
CENTS

# MAD, MAD MONEY

Mad, mad money for you to save or to spend,
PENNY, one cent; NICKEL, five; a DIME is ten.
A QUARTER, twenty-five cents; a HALF-DOLLAR, fifty;
A gold or silver DOLLAR COIN, which is real nifty.

There are also paper BILLS, or "cash," that help us to thrive:
Washington's on ONE dollar, and Abe Lincoln's on the FIVE,
Hamilton TEN, Jackson TWENTY, and Grant's on FIFTY—whew—
Ben Franklin's on ONE HUNDRED, and Jefferson's on the TWO.

| 25% | 50% | 75% | 100% |
|---|---|---|---|
|  |  |  |  |

# DISCOUNT PERCENTS

Percent is one way to show parts of a whole.
One hundred percent always equals it all,
Fifty percent equals one-half of one hundred,
Twenty-five percent, or half of fifty, discounts at the mall.

Discounts show a percent of the total price.
Keep your eyes on the sale prices at the store.
If you're ever at the mall and see 75 percent off,
Run there quickly because you'll rarely see more.

# FRACTION ATTRACTION

When do numbers have a fun attraction?
When they're used to form a type of fraction.
A fraction can show part of a SET or a WHOLE.
To understand all types of fractions is your goal.

The number on the top is the NUMERATOR.
The bottom number is the DENOMINATOR.
The top shows how often it occurs, or the frequency.
The bottom number is all the parts, or objects, totally.

Once you know that PROPER FRACTIONS equal less than a whole
And IMPROPER FRACTIONS equal one or more, you're on a roll.
Also, many fractions have more than one name,
Like one-half, two-fourths, and three-sixths equal the same.

You can ADD, SUBTRACT, DIVIDE, or MULTIPLY them,
COMPARE, ORDER, REDUCE, or even CONVERT them.
You can always use them on a NUMBER LINE.
Between the whole numbers they will brightly shine.

We have fun using different fractions every day,
When we COOK or DRINK or with the MUSIC we play.
Two-thirds cup of cocoa used to make the chocolate cake,
Play drums in three-fourths time, and share one-half a coke with Jake.

You can now understand the fraction attraction.
Using them each day will give you satisfaction.
You can enjoy them and have a positive reaction,
So take quick action now and have some fun with a fraction.

# FIVE'S GOT JIVE

The number five is very awesome and cool.
It's the age you start elementary school,
Kindergarten class for half a day,
Counting, drawing pictures, and some play.

Five minutes between each number on a clock,
Count by fives for minute time, tick tock, tick tock.
Please "take five," you have been working too hard,
A five-minute break to play in the yard.

Five toes on each foot, five fingers on each hand,
Many music groups are a five-person band,
Five players on one team's court of basketball,
Five days are in a school week, and that is all.

Abraham Lincoln's on the five-dollar bill.
Stars we draw have five points; real ones have nil.
All day long we use the number five:
"High five" and "Give me five." Five's got jive!

"TWEENS"
TEENS
17
18
ADOLESCENT
14
19
15
ADULT
COLLEGE
DRIVER'S LICENSE
16

# AWESOME TEENS

The most fabulous ages are the awesome teens,
They're even greater than the preteen years we call the "tweens."
After twelve and before twenty are the ages they're between,
The most interesting years anyone has ever seen.

Becoming an adolescent at the age of thirteen,
Getting your driver's license when you are sixteen,
Being your high school homecoming "king" or "queen,"
Then maturing to an adult when you become eighteen.

Getting a full-time job, your last two years of being a teen,
Or attending college to get a degree, which is really keen.
Saying "good-bye" to these special years after age nineteen,
And having lots of good memories, probably umpteen.

# COMPOSITE OR PRIME IS THE QUESTION

Is that number COMPOSITE or PRIME?
Kids ask that question all the time.
The answer is actually quite plain;
It's the number of FACTORS that you can name.

With FACTORS, all numbers have themselves and one.
If there are no more, then it's PRIME and you're done.
COMPOSITE can easily have some more,
Like twelve has twelve/one, six/two, and three/four.

Zero and one are not COMPOSITE or PRIME.
There are no exceptions at any time.
If you know the number's times facts, the answer's easy to get.
Just pull the factors from all its facts, and you are all set.

# CONSTANT NUMBERS

Numbers are *constant, organized,* and *clear.*
They are also *dependable* for you.
There's nothing about them that you need to fear,
Because numbers are very *logical* too.

You can ADD, SUBTRACT, DIVIDE, or MULTIPLY them,
You can use them in a RATIO or PERCENT,
You can GRAPH them or use them to MEASURE a hem,
You can use them to count MONEY to pay rent.

Right of point for DECIMAL, top and bottom for a FRACTION,
Locate a place or point with an ORDERED PAIR.
Numbers have a lot of uses you can put into action,
So start practicing all these ways if you dare.

# FUNKY NUMBER LINES

ZERO is in the middle of funky number lines,
With some numbers to the left and also to the right:
NEGATIVE left, less than zero, with small MINUS SIGNS,
POSITIVE right, more than zero, with no sign in sight.

WHOLE NUMBERS on the line, and there's sometimes more,
Like some FRACTION parts between them you might see—
For example, two-thirds between three and four,
Or one-half in the middle of two and three.

There are many ways you can use a funky number line,
Like ROUNDING numbers to a certain digit,
Or COMPARING sizes, like twelve is three more than nine,
And seeing the whole numbers and fractions between it.

# MY VERY FAVORITE NUMBER IS FOUR

My very favorite number is four,
And I have such good reasons galore.

I'll start by saying four's even and its shape is cute,
And number four has so many good uses, to boot.
First, there are the "Beatles," the famous "Fab Four,"
Or a quartet of four, no less and no more.

The bride, the groom, the maid of honor, and best man,
The number of wheels on a small truck, car, or van,
Just the right number of carrots for a snack,
Four lights on a car, two in front, two in back.

School is out when it's four o'clock,
And it's time to have some fun and rock.
A four-day camping trip is very cool,
Four years old and it's time for preschool.

In fourth grade, you say "bye" to your primary years.
After that, four more grades, and then high school with your peers.
Four years in high school and also in college,
So you can absorb a lot more knowledge.

Oh, number four, how I love you so,
More than any number, high or low.

# NUMBERS, NUMBERS EVERYWHERE

Numbers, numbers everywhere,
I use numbers all day long,
From when the sun rises up so bright
Till the sun goes down and is out of sight.

It all starts at 7:00, when my alarm startles and awakes me from my dream.
Then I quickly brush my teeth, get dressed, and comb my hair, so I don't miss bus 13.
It comes at exactly 7:30 and I cannot be late,
Because it only waits 1 minute for me and my brother, Nate.

But first I ask my mom for 2 pieces of toast
And 1 bowl of cereal that I like the most.
After breakfast, on the school bus, 21 kids I see,
And I know that Mr. Bumpers still needs to pick up 3.

Then we get to our school on 8 Pickle Drive,
And I walk to my 4th-grade class in room 5.
First Ms. Grape counts 3 absences and collects 10 dollars in milk money.
Then we discuss the day's weather forecast, which is 64 and sunny.

In math, we practice times facts, 1–12, so Friday's quiz we'll pass.
In science, we measure 7 grams for our 1 gummy worm's mass.
Next, during language arts, we discuss our reading of chapter 10.
Then I'm glad when it's lunchtime, and I share 1/2 my orange with Ben.

After lunch, in social studies, 12N/8W we locate a place.
Then my favorite class, PE, where I win by 5 seconds in a race.
Finally, it's 3:00, and it is time to board the school bus.
Mr. Bumpers stops at 9 Hops Road, and Mom makes snacks for us.

I gobble down 6 cheese crackers and grab the remote for TV.
Then I press channel 27 to watch a *Goosebumps Mystery.*
For dinner, I gobble 4 fish sticks and 1 piece of cake,
Then off to my soccer game and, hooray, 3 goals I make!

Last, I sit in bed and silently read 20 pages about Beckham's soccer team.
Then I fall asleep till 7:00, when my alarm startles and awakes me from my dream.
I wait in bed 1 minute as I yawn and rub my sleepy eyes.
Then I sit up and put my 2 slippers on, and out of bed I rise.

And then it starts all over, as I brush my teeth, get dressed, and comb my hair.
Then I notice 7:09 on my alarm clock, and I think and stare.
From dawn to dusk I am aware
That I see numbers everywhere!

# ODD MAUDE AND EVEN STEVEN

Odd Maude and Even Steven are good friends that disagree
About which numbers are used more, like *2* and *4* or *1* and *3*.
And even though their friendship is great,
They compare their numbers and debate.

*5* fingers, *5* toes, and *1* pink tongue,
*2* hands, *2* feet, and a *double* lung,
*1* nose, *1* mouth, and *1* red heart beneath,
*2* nostrils, *2* lips, and *32* teeth.

There are *triangles, pentagons, heptagons,*
*Quadrilaterals, hexagons, octagons,*
*Triplets, quintuplets, septuplets,*
*Twins, quadruplets, and octuplets.*

A piano has *25* keys that are shiny and black;
*36* keys are white, which is more if you're keeping track.
*3* push valves on a trumpet, *1* slider on a trombone,
*22* keys that you can push on a saxophone.

*9* defense players on the field in the game of baseball,
Ice hockey has *6* on the ice, who can possibly fall.
In tennis *singles*, there's 1 player on each side;
In *doubles*, *2* on each side try not to collide.

Maude and Steven can go on and on this way.
They can go on and on and on all day.
But they really need to stop and think and see
Odd and even numbers are used equally.

# HOORAY, ZERO

When I think about the number zero, I feel sad,
Because zero's lonely and worth nothing and feels bad.
You see, the number zero has no value on its own,
But right of a number, it's worth more than when it is alone.

It makes the number 10 times greater than the place before,
Like 5 tens and 0 ones is 50 or 10 times more,
Or 8 in the hundreds, 0 in the tens and ones
Makes it 10 times 80, and 800 it becomes.

The more zeros you put to the right of a number, the bigger it will be,
Like 100 zeros to the right of a one make it a googol, you see.
One of the ways that zero's special is it holds one place or more,
And that's how it makes the number bigger than it was before.

Another way that zero's special brings it to its fruition.
It has two properties with multiplication and addition.
The sum of zero and another number equals that number in addition.
The product of zero and a number equals zero in multiplication.

No other number, or digit, can ever do the same.
The Zero, or Identity Property, is its name.
Now I don't feel sad at all about the number zero,
'Cause it's special and important and a number hero.

# OPERATION VOCABULARY

Operation vocabulary is an important part of math.
It helps you choose the correct operation and follow the right path.
In addition, ADDEND plus ADDEND always equals the SUM.
Subtract MINUEND minus SUBTRAHEND, and the DIFFERENCE will come.
You get a PRODUCT when you multiply two FACTORS or so.
MULTIPLES are several PRODUCTS that you have in a row.
Divide a DIVIDEND by a DIVISOR, and a QUOTIENT you'll get.
Now you know the math operation vocabulary, and you're all set.

# DOES MEASUREMENT MEASURE UP?

Does measurement measure up? Well, let's just see.
There's LINEAR, WEIGHT, or MASS, and CAPACITY,
The three ways to measure in each system.
For US CUSTOMARY, let's now list them:
INCH, FOOT, YARD, and MILE are the LINEAR ones,
And WEIGHT is only OUNCE, POUND, and TONS.
There are many measures found in CAPACITY:
TEASPOON, TABLESPOON, and FLUID OUNCES are three.
The CUP, the PINT, the QUART, and GALLON are the rest.
To understand all these is your measurement quest.

For the METRIC system, you always start with the base.
Then go up or down three measures, whatever the case.
The three metric BASES are the METER, GRAM, and LITER,
Teamed with a PREFIX measure, this system is much neater.
The prefixes are always multiples of ten, so neat, no mess.
Ten, one hundred, one thousand times the base for greater; divide for less.
DECA, HECTO, and KILO are the prefixes larger,
DECI, CENTI, and MILLI are the ones that are smaller.
Each of the six prefixes branches off a base, or the main stem,
Eighteen measurements that you can use in the metric system.

So, does measurement measure up? I am sure that you can see
That measurement is totally awesome and always will be.
Whether customary or metric, it is quite plain.
They are both great to use, though they're not at all the same.

12
11
1
10
2
9
3
8
4
7
5
6
3:00
3:00
APRIL
SUNDAY
MONDAY
TUESDAY
WEDNESDAY
THURSDAY
FRIDAY
SATURDAY
1 2 3 4 5 6
7 8 9 10 11 12 13
14 15 16 17 18 19 20
21/28 22/29 23/30 24 25 26 27

# THE GIFT OF TIME

Time's an awesome gift we should appreciate,
HOURS, MINUTES, and SECONDS we cannot waste.
But rather, we should use it with great care,
Or save it and then have some time to spare.

Sixty seconds in a minute, sixty minutes in an hour.
The hour is the small hand, but it always has the power.
Seconds are the skinny hand, and minutes are the big one.
In a twenty-four–HOUR day, round and round the clock they run.

Seven DAYS, four WEEKS, a twelve-MONTH year and no more,
A ten-year DECADE, and twenty years is a SCORE.
A CENTURY's one hundred years, or a score times five, but then
A thousand-year MILLENIUM is a century times ten.

You can use CLOCKS and CALENDARS to help keep track of time.
Keep track and use time wisely is the message of this rhyme,
For time is very precious, and it is a gift for you.
Don't waste it, but always use it, and sometimes save it too.

25'

# COOL MATH TOOLS

There are many different types of cool math tools
That are used at many different types of schools.
A RULER, TAPE MEASURE, YARD, or METER stick—
These all measure length or distance in a lick.

Weight is measured in pounds and ounces on a WEIGHT SCALE.
A BALANCE SCALE using grams measures mass without fail.
A COMPASS draws and measures lots of circles, big and small.
Angles need PROTRACTORS to measure degrees of them all.

Use a TRUNDLE WHEEL to measure a perimeter.
CALIPERS for 3-D circles give diameter.
Using these cool math tools is helpful and fun.
They'll give you your measures, and then you'll be done.

"YES, it's CERTAIN!"
"NO, it's IMPOSSIBLE!"
"It's MORE LIKELY!"
"It's LESS LIKELY!"

# WHAT ARE THE CHANCES?

What are the CHANCES a certain OUTCOME will be
CERTAIN, IMPOSSIBLE, or more or less LIKELY?
What are the CHANCES? Are they PROBABLE or not?
It depends on the LIKELINESS in the EVENT you've got.
What are the CHANCES your PREDICTION will be right?
In the EXPERIMENT, it may not, but it might.
What are the CHANCES you'll take a RISK, high or low?
You MIGHT say "yes" but you will PROBABLY say "no."
What are the CHANCES? Are they good, or are they bad?
They PROBABLY could be either, so don't feel sad.
What are the CHANCES you'll finish this poem that is fun?
The CHANCES are CERTAIN, because now you are done!

# 3 M'S AND AN R

The MODE, the MEDIAN, the MEAN, and the RANGE,
The three M's and one R that never change.
They help you learn about your collected facts,
So there's nothing about your data that lacks.

The number that occurs most often is the MODE;
In the whole set of numbers, it carries the load.
The MEDIAN is the middle number of the entire set;
When the numbers are in order, it is easier to get.

You divide up the sum to get the third M, called MEAN,
And you end up with the average, and that's real keen.
The only R is the RANGE of it all,
The spread of the data from large to small.

# A WHOLE LOT OF GRAPHS

There are a lot of pictures with data, called graphs.
Some have BARS, some have PICTURES, some have LINES,
Often placed on a GRID with whole numbers and halves,
Some have a KEY with symbols and signs.

Many types of facts can be shown on these graphs.
They are quite versatile, you will find,
Like comparing the heights of buildings and giraffes,
Using one, or double, bars or lines.

Another is a CIRCLE divided into parts,
Used to show all the pieces of a whole.
Place value numbers on STEM-AND-LEAF PLOT charts,
X's for frequency on LINE PLOTS is the goal.

# COORDINATE YOUR COORDINATE GRID

Coordinate you coordinate grid with the AXES X and Y.
Draw X HORIZONTAL and Y VERTICAL—now give it a try!
Put LINES between the axes, NUMBERS on the bottom and side.
Write ZERO in the corner of the grid, where it will reside.
And please don't forget the TITLE, that tells what it's about.
Write it on top in big, dark letters to make it stand out.
Use an ORDERED PAIR to draw or locate a point or a place.
First number, X RIGHT, then second, Y UP, with zero as your base.

# A CIRCULAR CIRCLE

A circular circle has points the same distance from the middle.
Round and round the points all go as they curve round just a little.
The circle's CENTER POINT letter always gives the circle its name;
Beginning a RADIUS, or half diameter, it does the same.

A segment from one outer point to another is a CHORD.
A DIAMETER's the same, but through the center it has soared.
CIRCUMFERENCE is the measurement around the whole outside;
It never measures anything at all on the inside.

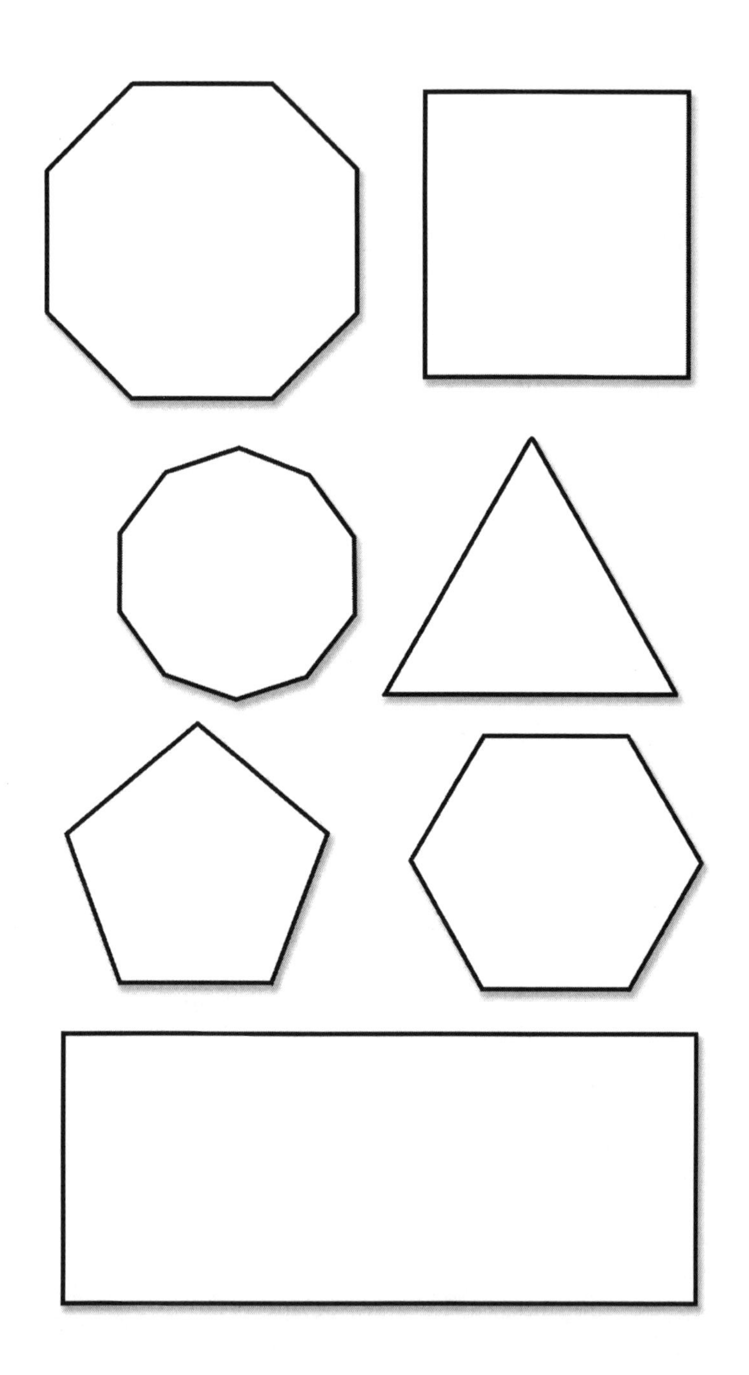

# POLYGONS

Polygons are closed, and they have straight sides,
And inside every corner an angle hides.
They can have three or many sides and be topsy-turvy,
But no part of a polygon can ever be curvy.

# AREA HYSTERIA

If you don't understand how to measure a *rectangle's area*,
It can surely end up causing you some great hysteria.

There are many wrong measurements that you might get
For your inside rug or outside pool, you can bet.

First, the mathematic formula is what you need to know:
Area of rectangles equals length times width, but stop now—whoa!

You also need to understand its unit of measure:
They are always square, and finding them will give you pleasure.

A square unit example is an inch by inch by inch by inch,
And now there's no hysteria, because you see that it's a cinch.

If you ever need a different polygon's area,
Learn its special formula to avoid hysteria.

# POLYGONS, POLYGONS EVERYWHERE

Polygons, polygons everywhere!
We see them when we are aware
Of our surroundings every day,
When we are looking every way.

A beautiful RECTANGULAR picture frame,
A SQUARE board for a Monopoly game,
A DIAMOND shape is the baseball field,
A yellow TRIANGULAR sign means "yield."

A bright red OCTAGON on the road says "stop,"
The PENTAGON, with five sides but one roof on top,
A RHOMBUS here, a TRAPEZOID there,
You can find polygons everywhere!

# POLYGONS, POLYGONS, I LOVE YOU

Polygons, polygons, I love you.
TRIANGLES and QUADRILATERALS just name two,
PENTAGONS and HEXAGONS now make four,
HEPTAGONS and OCTAGONS are two more,
NONAGONS and DECAGONS will make eight.
Having these different polygons is so great.
The smallest number of sides they have is three,
And the largest number is infinity.
Polygons can have many sides galore,
'Cause with sides you can always add one more.

# RELATE TO LINE RELATIONSHIPS

There are two main line relationships to which you can relate.
PARALLELS are side by side and never touch 'cause it's their fate.

But INTERSECTING lines do touch, because they cross each other;
These lines can touch right now or cross later with one another.

PERPENDICULAR are special types of intersecting lines;
They cross and create four right angles, which is always their design.

There's one more type of intersecting line that you need to know:
TRANSVERAL lines pass through other lines, cutting them in a row.

# THE RIM OF PERIMETER

What is a perimeter measurement? I ask you.
Well, the word "rim" in perimeter is your big clue.
It is the total measurement of any polygon's rim.
Just add up all the side measures on the outside, not within.

But a rectangle is special, and its formula is this:
Two times length plus two times width, and you can't miss.
And a square has a specific formula too:
It's four times the measurement of one side – it's true!

When do you need the measurement of a polygon's rim?
When you use a tape measure round a room for total trim.
Another is when you put a fence around your yard;
You can use a trundle wheel to help, and it's not hard.

# THREE MAIN ANGLES

The number of main angles in geometry is three.
The first is squared off sharply and named RIGHT.
This angle's measure always equals ninety degrees.
Protractors measure the angle's width, not the height.

The second is a big angle, which is called OBTUSE.
More than ninety, less than 180 it will be.
It measures between these two numbers, so you can deduce
That it is greater than a right angle, you see.

The last is the very cutest angle of them all.
It's less than ninety and more than a straight line.
It is called ACUTE, and it can be very small,
Or it can even measure up to eighty-nine.

# TRICKY TRANSFORMATIONS

There are tricky transformations you can use with a flat shape
To move a figure different ways and quickly change its scape.
The first is a TRANSLATION, also known as a SLIDE.
Move diagonal, up, down, right, or left, and take it for a ride.

Next is a REFLECTION, also called a FLIP transformation.
When shapes flip over a line, make a close observation,
Because a mirror image the reflection will always make.
It looks the same but backwards, and it's no mistake.

The third move is called a TURN, also known as a ROTATION.
The figure moves around a point in a circle formation.
One-fourth, one-half, and three-quarters are each a different size turn.
The shape must always look the same, which is important to learn.

# VOLUMES ABOUT VOLUME

Volumes about volume, you certainly do not ever need to read
If it's the volume of a *cube* or *rectangular prism*, which are 3-D.
The formula and measure unit are all you need to know,
And to really understand these both is your important goal.
The formula has three parts, 'cause it has a third dimension,
So you multiply together, length times width and then times height.
Volume's cubic unit is important for your retention.
Using this formula and measure, you will always get it right!

"hmmmmmm.......
This makes SENSE!!"

# IS LOGIC LOGICAL?

Is logic logical, and does it make sense?
Sometimes it is fun, but other times intense.
It helps you ELIMINATE and NARROW CHOICES down.
It helps you FIND YOUR ANSWER, so that you do not frown.
Yes, logic is logical, because it's REASONABLE THINKING.
It makes a lot of sense, and it keeps your THOUGHTS LINKING.

# A VARIETY OF VARIABLES

There's a variety of variables that you might see
In algebraic expressions and equations, if need be,
Like 4+x, 8-n, or 2+y=3:
They're letters that stand for missing numbers, or amounts, you see.
Finding what the variable equals is always the key,
Because it helps you get your answer, and that's a guarantee.